Air Power

F/A-18 SUPER HORNET

MEGAN COOLEY PETERSON

BLACK RABBIT BOOKS

Bolt is published by Black Rabbit Books
P.O. Box 227, Mankato, Minnesota, 56002.
www.blackrabbitbooks.com

Jennifer Besel & Marysa Storm, editors;
Grant Gould, designer; Omay Ayres,
photo researcher

Library of Congress Cataloging-in-Publication Data
Names: Peterson, Megan Cooley, author.
Title: F/A-18 Super Hornet / by Megan Cooley Peterson.
Description: Mankato, Minnesota : Black Rabbit Books, [2020] | Includes bibliographical references and index. | Audience: Grades 4-6. | Audience: Ages 8-12.
Identifiers: LCCN 2018031984 (print) | LCCN 2018033219 (ebook) | ISBN 9781680727944 (e-book) | ISBN 9781680727883 (library binding) | ISBN 9781644660119 (paperback)
Subjects: LCSH: Hornet (Jet fighter plane)—Juvenile literature.
Classification: LCC UG1242.F5 (ebook) | LCC UG1242.F5 P4735 2020 (print) | DDC 623.74/63—dc23
LC record available at https://lccn.loc.gov/2018031984

Printed in the United States.

Image Credits

af.mil: Staff Sgt. Shawn Nickel/US Air Force, 22; afcent.af.mil: Staff Sgt. Aaron Allmon/ US Air Force, 20–21; aiirsource.com: Aiir Source Military, 25; airforce.gov.au: Sgt Pete/Australian Defence Force, 13 (b); Alamy: Stocktrek Images, Inc., 9-Jan; amc.af.mil: Senior Airman Madelyn Brown / US Air Force, Cover; boeingimages.com, Boeing, 4–5; commons.wikimedia.org/af.mil: ROB TABOR/US Air Force, 12 (b), 31; commons.wikimedia.org/defenseimagery.mil: U.S. Marine Corps Capt. Christopher Prout, 6–7; coub.com: Grange, 3; flickr.com: Airman 1st Class Brett Clashman/US Air Force/Nellis Airforce Base, 10 (b); Getty: Ian Hitchcock, 19 (t); navair.navy.mil: NAWCTSD, 26–27; navy.mil: Mass Communication Specialist 2nd Class Z.A. Landers/US Navy, 28–29; Mass Communication Specialist 3rd Class Matt Matlage/US Navy, 26 (t); Mass Communication Specialist Seaman Andrew Schneider/US Navy, 27 (b); Mass Communication Specialist Seaman Pasquale Sena/US Navy, 19 (b); Seaman Christopher A. Michaels/US Navy, 13 (t); Shutterstock: Anatoliy Lukich, 1; Benguhan, 12–13 (bkgd); Chris Parypa Photography, 15; Eugene Berman, 32; fusebulb, 10 (t); inspired-fiona, 16 (bkgd); the-blueprints: the Blueprints, 16–17

CONTENTS

CHAPTER 1

Taking Down the

It was June 2017. Soldiers were fighting enemies in the country of Syria. Suddenly, an enemy fighter jet dropped bombs on the soldiers. Other aircraft warned it to stop. A U.S. F/A-18 Super Hornet zipped to the scene.

The Super Hornet's mission took about eight minutes.

Mission Complete

The enemy jet ignored the warnings. The Super Hornet shot two **missiles** at it. One missile hit its target. The jet crashed to the ground.

CHAPTER 2

HISTORY of the F/A-18 Super Hornet

The Super Hornet's history started in the 1980s. The U.S. Navy started using a new fighter jet called the F/A-18 Hornet. It took off from **aircraft carriers**. It fought air-to-air missions. It also bombed ground targets.

104

The Super Hornet

is nicknamed the "Rhino."

Super Hornet	vs.	Hornet
1 or 2	crew	1 or 2
1,458 miles (2,346 kilometers)	distance it can go before refueling (at combat speed)	1,252 miles (2,015 km)
more than 1,381 miles (2,223 km) per hour	top speed	more than 1,304 miles (2,099 km) per hour

Improving the Hornet

By the 1990s, the Hornet needed updating. The updated jet was called the Super Hornet. It could fly longer and farther. It also carried more weapons.

SUPER HORNET TIMELINE

1999
The jet enters active duty.

1990

November 29, 1995
The Super Hornet makes its first flight.

2018
The U.S. Navy orders upgrades to the jets.

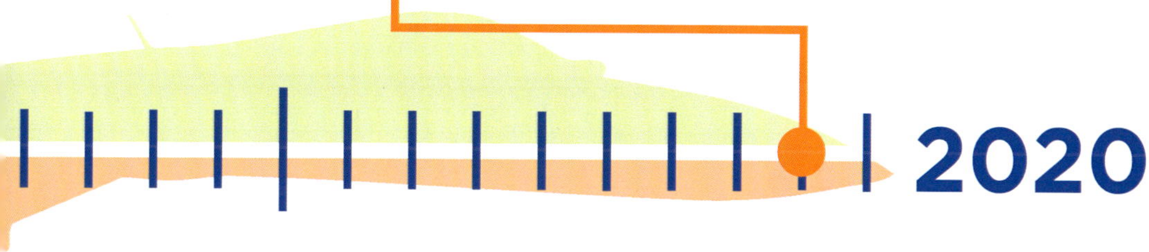

2002
Super Hornets fly missions in Iraq.

CHAPTER 3

Super Hornet FEATURES

The Super Hornet soars at more than 1,381 miles (2,223 km) per hour. Two engines create 44,000 pounds (19,958 kilograms) of thrust. Thrust pushes the jet through the air.

TOP SPEED COMPARISONS

more than 1,381 miles (2,223 km) per hour
1,200 miles (1,931 km) per hour
about 1,500 miles (2,414 km) per hour
1,000
1,100
1,200
1,300
1,400
1,500
1,600

STATS

wingspan
44.7 feet (13.6 meters)

height
16 feet (4.9 m)

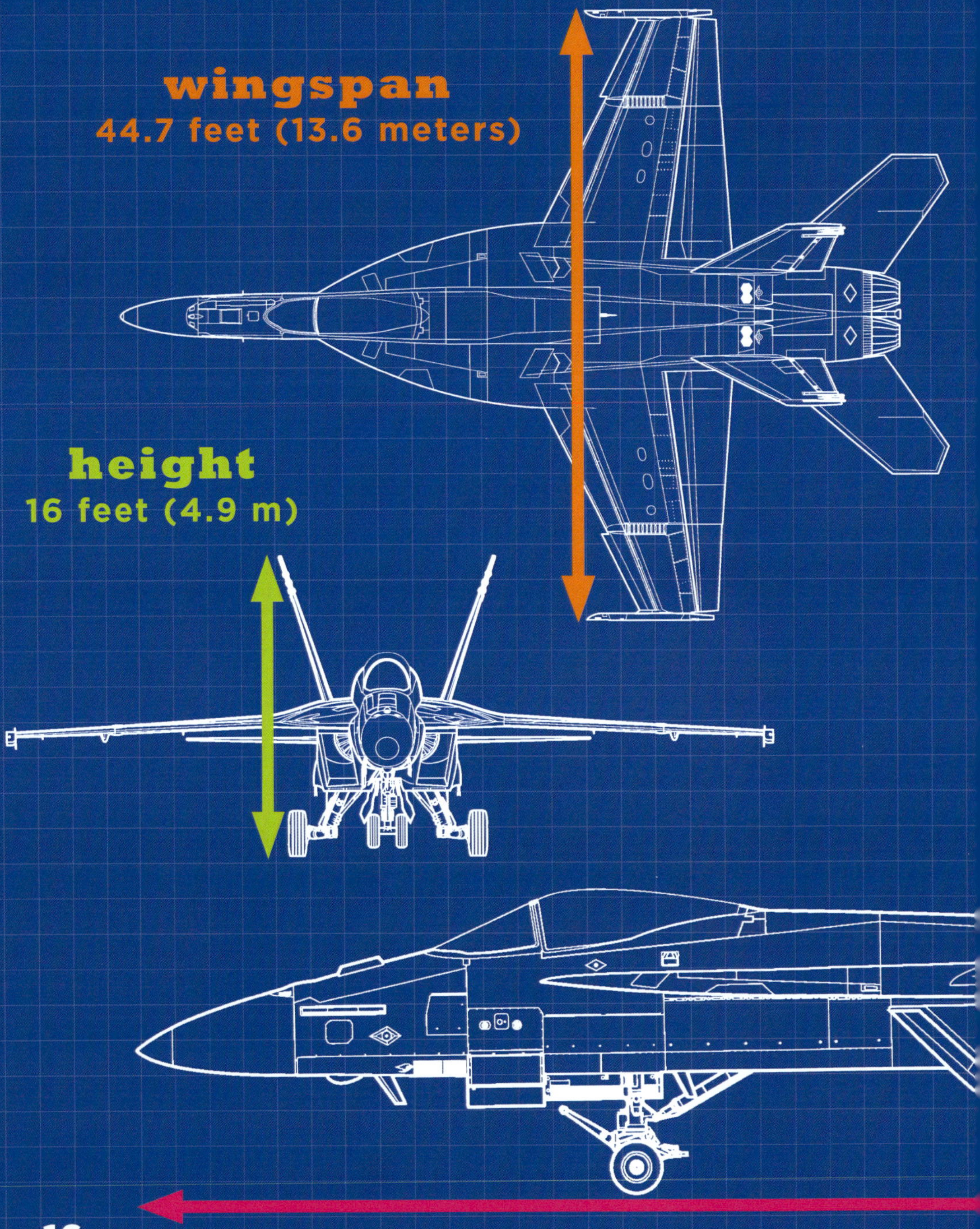

Fuel

The Super Hornet carries more than 29,000 pounds (13,154 kg) of fuel. Having this much fuel is helpful to other aircraft. The Hornet can fill up other planes in the air.

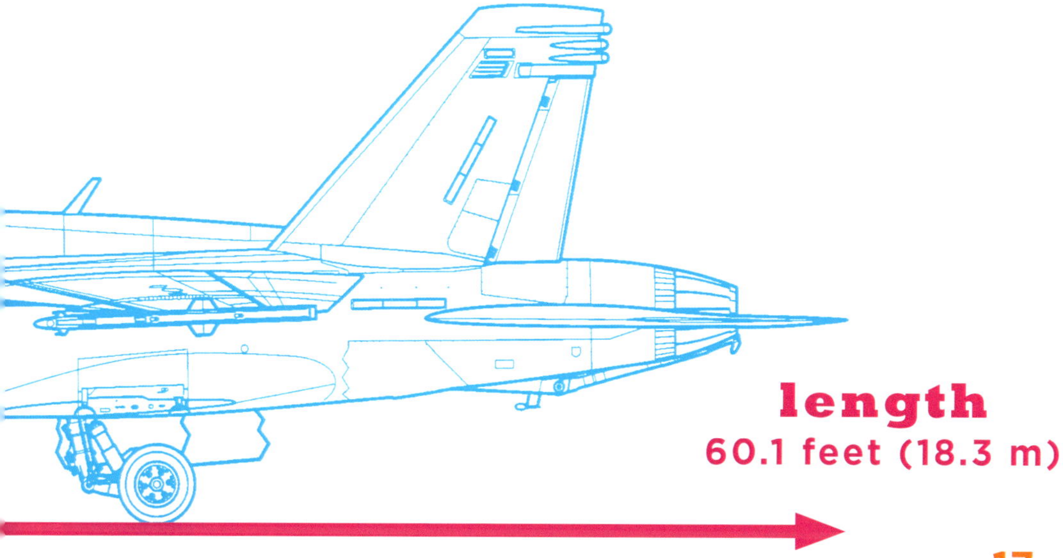

Weapons

The Super Hornet is used for air and ground missions. It shoots missiles at other aircraft. It also drops bombs on ground targets. Its **Gatling gun** fires up to 6,000 rounds a minute.

Australia uses Super Hornets too.

SUPER HORNET PARTS

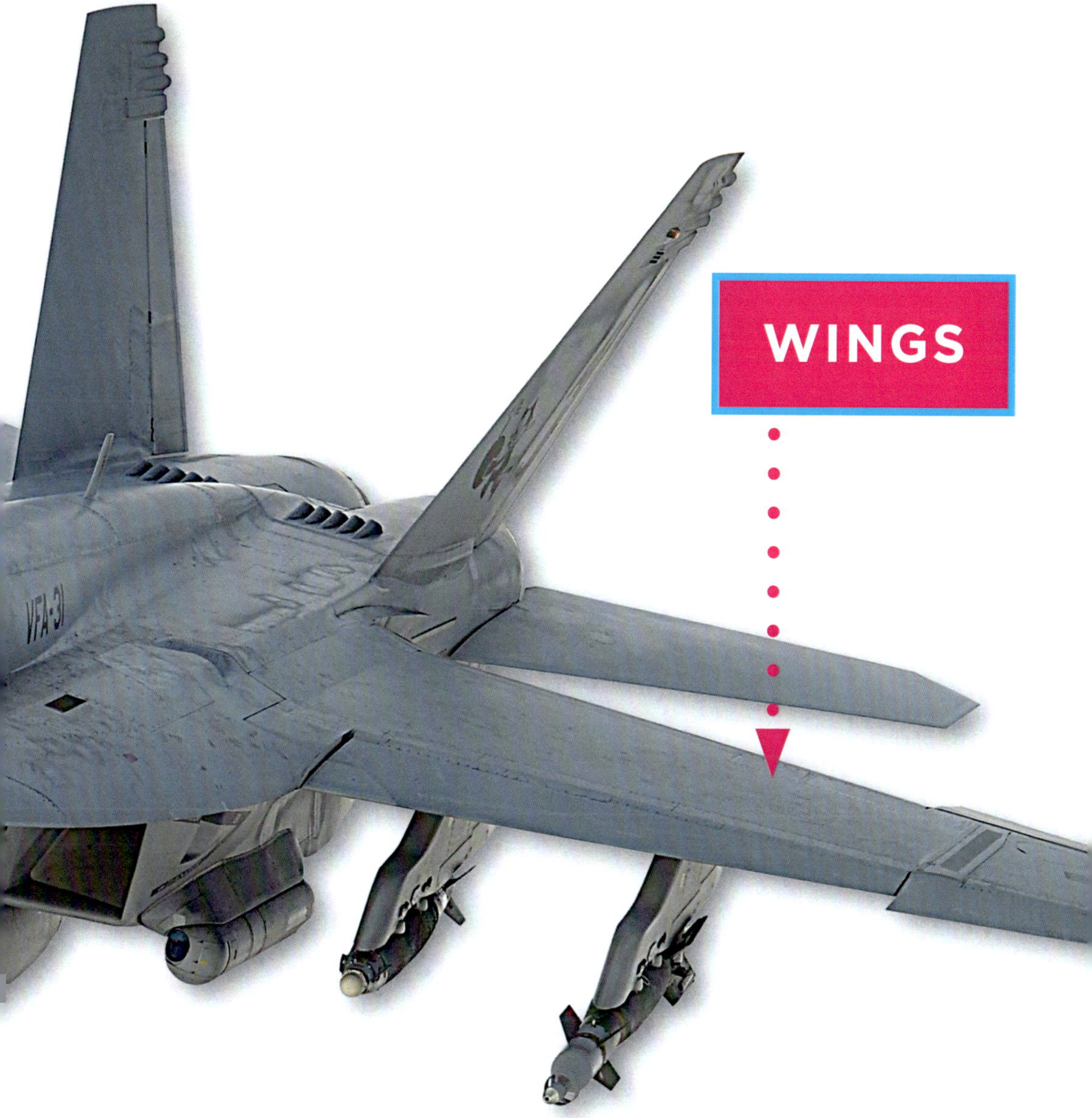
VFA-31
WINGS
FUEL TANKS

CHAPTER 4

The Super Hornet in ACTION

The sight of a Super Hornet scares away many enemies. In 2002, the jets began flying over Iraq. They made sure enemy aircraft didn't enter **no-fly zones**.

Nowhere to Hide

Stealth aircraft often slip past **radar**. But the Super Hornet doesn't miss them. Its **sensors** search for heat aircraft leave behind. It uses that heat to find enemies.

BY THE NUMBERS

11 NUMBER OF WEAPONS STATIONS ON EACH SUPER HORNET

more than 50,000 feet (15,240 m)
HOW HIGH A SUPER HORNET CAN FLY

$57 MILLION cost to buy one Super Hornet

66,000
pounds
(29,937 kg)
MAXIMUM
WEIGHT AT

401

Ready to Help

The Super Hornet is a trusted fighter. It keeps ships, soldiers, and other aircraft safe. The jet will fly into action for years to come.

The U.S. military has planned upgrades for the jets. The new Hornets will have more stealth and better computers.

GLOSSARY

aircraft carrier (AIR-kraft KAHR-ee-uhr)—a warship with a deck on which aircraft can be launched and landed

Gatling gun (GAT-ling GUN)—a machine gun with a spinning cluster of barrels

missile (MIS-uhl)—a weapon that can be thrown or projected to hit an object far away

no-fly zone (NO-FLY ZOHN)—an area where military planes are not allowed to fly during a war

radar (RAY-dar)—a device that sends out radio waves for finding the location and speed of a moving object

sensor (SEN-sor)—a device that finds heat, light, sound, motion, or other things

stealth (STELTH)—an aircraft design that is hard for radar to pick up

BOOKS

Harris, Tim. *Superfast Jets: From Fighter Jets to Turbo Jets*. Feats of Flight. Minneapolis: Hungry Tomato, 2018.

Kennedy, Robert. *Would You Dare Fly a Fighter Jet?* Would You Dare? New York: Gareth Stevens Publishing, 2017.

Sherman, Jill. *U.S. Navy*. Serving Our Country. Mankato, MN: Amicus, 2019.

WEBSITES

F-18 Super Hornet Demo at Aviation Nation 2016
www.youtube.com/watch?v=p8SJ98CEqsQ

F/A-18 Hornet Strike Fighter
www.navy.mil/navydata/fact_display.asp?cid=1100&tid=1200&ct=1

F/A-18 Super Hornet
www.boeing.com/defense/fa-18-super-hornet/

INDEX